50 STEAM PROJECTS FOR KIDS SAFE ENOUGH TO TASTE

5-MINUTE KITCHEN EXPERIMENTS

WOLF CUB CHLO

Free Gift

Thank you for your purchase!

to download your:

- free colorful science journal for your junior scientists
- 10 additional fun and quick kitchen science experiments

To all our scientists young and mature, keep dreaming, keep reaching, keep reading and keep having fun!

Contents

SCIENCE

TECHNOLOGY

ENGINEERING

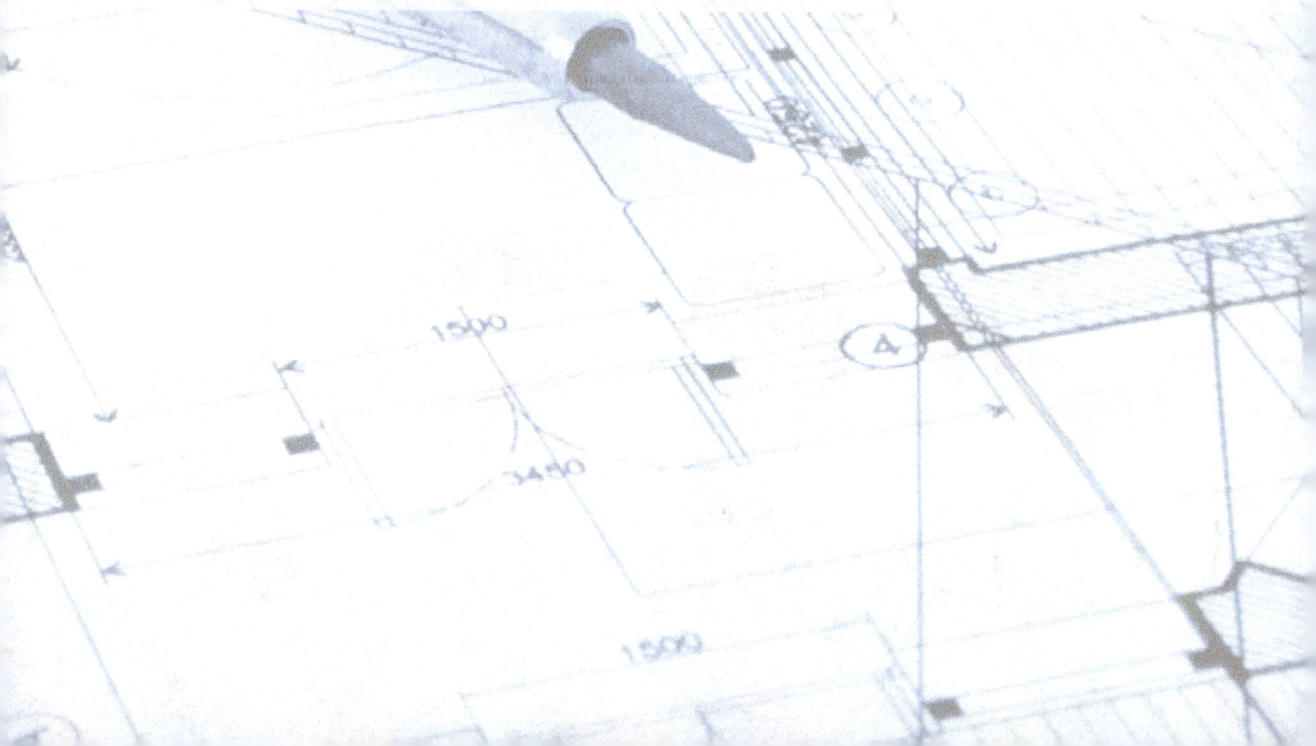

ART

MATH

NOTE TO ADULT SCIENTISTS

Hey there, fellow grown-up scientists. Science is fun. Science is messy but some experiments can be difficult to set up, time-consuming, and toxic.

The goal of this book is to make sure that all experiments are not only fun, but fast, non-toxic, safe to taste, and easily accessible, with most ingredients found right in your kitchen.

No more worrying about:

- your mini scientist ingesting chemicals.
- wasting time and money running to the nearest grocery store to gather a long list of hard-to-find items.
- Spending more time on setting up the experiment than conducting the actual experiment itself.

Instead, take pleasure in spending time with your little scientists, helping them learn while having fun with these simple, inexpensive science experiments while encouraging them to play with ingredients found right in your pantry.

Each experiment has an Edible meter: **non-toxic** means food is safe to touch and play with but not to eat, **Taste safe** means the experiment is safe to taste, but not eat entirely and **edible** means the experiment is safe enough to eat entirely.

Every receipe can be categorized into each **STEAM** category, but we have divided the experiments into sections so they can be easier to locate.

STEAM can be incorporated into any experiment by doing the following:

Science: Discussing how certain particles or chemical agents interact with each other.

Technology - Taking pictures of each experiment automatically incorporates technology. Uploading, sending and sharing the projects are extra bonus points. Creating/using tools to solve everyday problems also counts.

Engineering - Will the experiment have the same results if the containers are enlarged or made smaller? Does size and shape affect the outcome? Which type of engineering is being explored? How many types can you incorporate into the project?

Art - Adding food coloring to the experiments and drawing pictures adds the artistic element.

Math - Measuring additives, changing measurements, omitting and adding different ingredients always changes results. Try it out for fun.

Please note, if an experiment is edible but extremely sugary, I still wouldn't recommend eating it all, but what you do in your kitchen lab, is entirely up to you and your little scientists.

Surprise Free Gift: Your Free gift is a colorful science journal for your junior scientists to document all their questions, observations and conclusions and 10 additional fun and quick kitchen science recipes. The Science Journal encourages advanced children to document their analyses and hypotheses; while encouraging younger children to express their analyses and observations through drawings.

Warm scientific regards,
Mama Wolf

NOTE TO KID SCIENTISTS

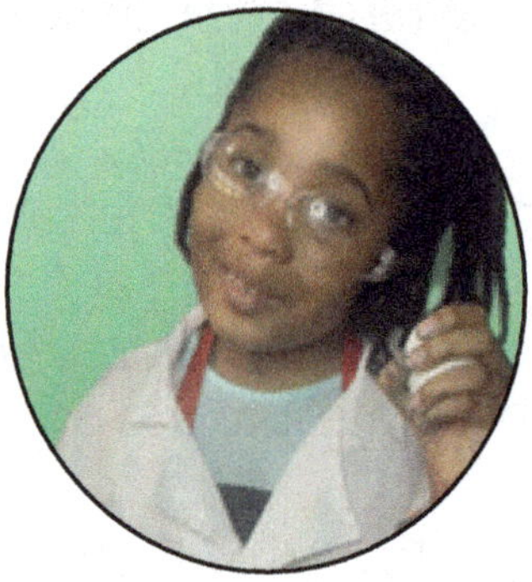

Hi, Junior scientists! You are finally allowed to play with everything found right in the kitchen! Good scientists always ask questions and take notes.

Remember to always read directions with a grown-up so you can stay safe in the kitchen. But, most importantly, remember to explore and have some fun!

Wolf Cub Chlo

What's a scientist's favorite candy?
Experi-mints!
100ml
APPROX
80
100 ml

SCIENCE

Science is the big topic you will be using to explore every experiment. Exploring science is like doing detective work and discovering mysteries. Every scientists asks questions, makes observations and documents their conclusions and you will be expected to do the same as you explore the world behind the science of items found right in your kitchen.

In the next 12 experiments, you will investigate scientific ideas that focus on chemical reactions that cause a rush of mini bubbles and liquid to come skyrocketing out such as - Fizzy ice, Apple Volcanoes, Lemon Volcanoes and the colorful Gelatin Eruption. Ever wanted to learn how to build your own edible gummy bear? You will have the entire procedure right at your fingertips.

You'll discover how to bounce an egg, and even create your own slime out of play dough.

This chapter also answer questions such as.

How to make corn kernels dance?

Can you really trick your tastebuds?

How can soda be used to blow up a balloon?

And so much more! So let's get this kitchen lab all set up and ready to go!

PUDDING PLAYDOUGH SLIME

EDIBLE METER: Taste safe

TOOLS: Small saucepan

VOCABULARY: Non-newtonian, polymer

QUESTION: How does play dough slime feel ?

INGREDIENTS

- ¼ cup instant pudding
- 1 cup of corn starch (divided)
- 1/2 cup warm water (divided)

PROCEDURE

1. Mix instant pudding, water, and cornstarch in a bowl
2. Slowly add other half of cornstarch to the mixture
3. Slowly pour in water and observe the mixture while kneading by hand

Observations: Is it a solid or a liquid?

Note:

Store in an airtight container; discard after 1-2 uses

Fun Science Fact:

All slimes are non-newtonian liquids. This means they don't have a shape, just like liquids but when you touch them they feel solid. The pudding contains gelatin which is a polymer and so is cornstarch, so, when they are combined they create a cross polymer which makes the gooey texture.

BUILD A GUMMY BEAR

EDIBLE METER: Edible

TOOLS: Saucepan, silicone molds

VOCABULARY: reversible change

QUESTION: How can you form a solid from a liquid?

INGREDIENTS

- ½ cup cold water
- ¼ cup corn syrup
- 1 packet of flavored gelatin dessert powder
- 2 plain gelatin packets

Note:

To preserve, store them in an airtight container.

PROCEDURE

Mix water and corn syrup until clear and add to saucepan

Add a packet of gelatin, 2 packets of gelatin powder to the mixture in the saucepan, and whisk.

Caution: Hot! Turn the stove on medium-low heat and continue to stir until the solution bubbles, then pour into a bowl and let sit for a few minutes.

Caution: Hot! Scrape froth off the top of the mixture and pour carefully into molds.

Let the mixture settle for approximately 30 minutes; then transfer to the refrigerator for 2 hours.

Pop gummies out of molds and Enjoy!

Fun Science Fact:

Creating gummy bears is a chemical process of turning a liquid containing gelatin into a solid. If you put your gummy bears into the microwave, the gelatin will cause the gummy bears to turn back into a liquid. This process is called thermoreversible or reversible change.

WHAT'S POPPING? (CARBONATED CANDY)

EDIBLE METER: Edible

TOOLS: Funnel, (Optional: What's Popping? Observation worksheet)

VOCABULARY: carbon dioxide

QUESTION: How can soda be used to blow up a balloon?

INGREDIENTS

- 4 packets of carbonated candy
- 4 20 oz bottles of soda
- 4 balloons

PROCEDURE

Use a funnel to pour an entire packet of carbonated candy into each balloon.

Remove the tops on each bottle and carefully place balloons over the opening. (Note: Careful not to pour candy inside the bottles yet.)

Next, flip each balloon over and observe what happens as the candy drops into each bottle.

Fun Science Fact:

The secret to carbonated candy "popping" is that they have carbon dioxide bubbles inside of them. Soda also contains carbon dioxide; when the two contents mix, the carbon dioxide tries to escape and rises up, filling the balloon.

Nutrition Facts
Serving size
Amount per serving
Calories
Total Fat 0g
Sodium 45mg
Total Carbohydrate 58g
Total Sugars 58g
Protein 0g
210 CALORIES
16.9 FL OZ
(1.05 PT)
500 mL
POP ROCKS
POP ROCKS
POP ROCKS
POP ROCKS
CA CRV
Nutrition Facts

APPLE VOLCANOES

EDIBLE METER: Taste safe

TOOLS: tray, spoon, optional: pipette/dropper, (optional: apple volcanoes observation worksheet)

VOCABULARY: chemical reaction, base, acid, gas, carbon dioxide

QUESTION: What happens when baking soda and vinegar are mixed?

INGREDIENTS

- 3 apples
- Food coloring
- Vinegar
- Baking soda

PROCEDURE

Cut a hole on the top of the apple

Pour a spoonful of baking soda into the apple. (Note: Adding dish soap creates more bubbles to the reaction, but it will no longer taste safe.)

Add 5-6 drops of food coloring inside each apple.

Slowly pour about a tablespoon of vinegar into each apple and observe the reaction. (Fun tip: Mix colors to create new colors during the chemical reaction)

Fun Science Fact:

Explore chemistry right from your own kitchen! When baking soda (base) is mixed with an acid (vinegar) they create a gas called carbon dioxide which creates an explosive reaction

BOUNCY EGG

EDIBLE METER: Non-toxic, but not taste safe

TOOLS: jar, (Optional: Colorful Rubber Egg worksheet)

VOCABULARY: absorption, carbon dioxide

QUESTION: How to dissolve an egg shell?

INGREDIENTS

- Eggs
- Glass jar
- White vinegar (optional: food coloring)

PROCEDURE

Pour 1 cup of white vinegar in a jar and add 4-5 drops of food coloring

Carefully place the egg inside the jar. (Note: make sure the entire egg is submerged in the vinegar. You will need a separate egg for each color.)

Use the Colorful Rubber Egg worksheet to document your observations of the egg for the next 3 days. Remove on the 3rd day and rinse with water.

Take turns dropping eggs from shorter distances. For a major explosion, drop the egg from a higher elevation. (Caution: make sure to wash your hands after playing with the egg)

Fun Science Fact:

Eggshells are made of a chemical called calcium carbonate. When the eggshell absorbs the acid in the vinegar, it creates carbon dioxide which starts to dissolve the shell. Once the shell is dissolved, the egg absorbs vinegar and begins to "grow' in size.

FIZZY ICE

EDIBLE METER: Taste safe

TOOLS: ice tray, large bin/tray, pipette

VOCABULARY: base, acid, carbon dioxide (optional: Fizzy Ice Cubes worksheet)

QUESTION: What happens when vinegar is added to frozen baking soda?

INGREDIENTS

- Water
- Food coloring
- Baking soda
- Vinegar

PROCEDURE

Pour approximate 1tsp of baking soda into each cup inside an ice tray

Add food coloring diluted with water to each cup and freeze.

Remove cubes from the tray. Use a pipette to drop vinegar on cubes and observe the reaction.

Fun Science Fact:

Baking soda (base) and vinegar (acid) create a chemical reaction by creating carbon dioxide (CO_2) when mixed.

CELERY COLORS

EDIBLE METER: Edible

TOOLS: jars, (optional: celery colors worksheet)

VOCABULARY: osmosis

QUESTION: How do plants absorb water?

INGREDIENTS

- Celery stalks
- Food coloring
- Water

PROCEDURE

Fill jars halfway with water and add 15-20 drops of food coloring. (Note: You will need a separate jar for each color.)

Observe the celery stalks in 1-4 days. Document your observations

Fun Science Fact:

Plants need water to remain firm and grow. The process in which water travels through a plant is called osmosis. This process is usually invisible to humans, but adding food coloring helps you to see the process.

GELATIN ERUPTION

EDIBLE METER: Taste-Safe

TOOLS: small cups/bowls, large dish/tray

VOCABULARY: base, acid, chemical reaction, carbon dioxide (CO2)

QUESTION: What happens when gelatin dessert packets are mixed with baking soda and vinegar?

INGREDIENTS

- Gelatin dessert packets (a separate packet for each color)
- 2 tbsp baking soda
- ½ cup vinegar

PROCEDURE

Place small cups inside the dish tray or bin to contain any overflow of liquids.

Add 2 tbsp of baking soda to small bowls.

Mix 2-3 tbsp of gelatin powder with baking soda.

Pour vinegar into the bottle and document observations. What do you see? What do you hear? What do you smell? How does this experiment smell different from the others?

Fun Science Fact:

Baking soda (base) and vinegar (acid) create a chemical reaction by creating carbon dioxide (CO2) when mixed. The bubbles from the carbon dioxide help the kernels "dance."

BAKING SODA BALLOON

EDIBLE METER: Non-Toxic

TOOLS: water bottle, balloon, (optional: funnel, "Baking Soda Balloons" observation worksheet)

VOCABULARY: base, acid, chemical reaction, carbon dioxide (CO2)

QUESTION: Can a balloon be blown up without a person or machine?

INGREDIENTS

- 2 tsp baking soda
- Vinegar

PROCEDURE

Fill water bottle halfway with vinegar.

Stretch balloon before use. Use a funnel to add 2 tsp of baking soda into the balloon and attach it to the water bottle.

Lift the balloon to empty the baking soda into the bottle and observe the reaction.

Fun Science Fact:

Baking soda (base) and vinegar (acid) create a chemical reaction by creating carbon dioxide (CO2) when mixed. Carbon dioxide tries to escape; since it has nowhere to go, it rises to the top and fills the balloon. When you blow into a balloon, you are blowing out carbon dioxide to fill the balloon too.

CORN DANCE

EDIBLE METER: Taste Safe

TOOLS: water bottle/jars, (optional: funnel, "Dancing Corn" worksheet)

VOCABULARY: base, acid, chemical reaction, carbon dioxide (CO2)

QUESTION: What happens when you add corn to baking soda and vinegar?

INGREDIENTS

- ¼ c. corn kernels
- 2 tsp. baking soda
- ½ c. vinegar
- 2 cups water (Optional: food coloring)

FUN!

Fun Tip:

Try with clear club soda instead of baking soda and vinegar and document your results.

PROCEDURE

Fill a water bottle with 2 cups of water.

Mix 2 tsp of baking soda into the water.

Document what you think will happen to the kernels once the vinegar is added.

Pour corn kernels into a jar.

Slowly add ½ cup of vinegar to the bottle and document observations.

Fun Science Fact:

Baking soda (base) and vinegar (acid) create a chemical reaction by creating carbon dioxide (CO2) when mixed. The bubbles from the carbon dioxide help the kernels "dance."

LEMON VOLCANO

EDIBLE METER: Taste safe

TOOLS: Large tray/bin, fork, (optional: lemon volcano worksheet)

VOCABULARY: base, acid, chemical reaction, carbon dioxide (CO2)

QUESTION: What happens when you mix baking soda and lemons?

INGREDIENTS

- Lemon halves
- Baking soda
- Vinegar
- Food coloring

Note:

dish soap can also be used for an extra bubbly eruption, but then it will be no longer taste safe

PROCEDURE

Roll lemons to open up the juices and cut them in half. Pierce the flesh with a fork to open up the juices even more. Slice the bottom of the lemons so they can stand upright in the tray/bin.

Add several drops of food coloring to each lemon.

Dump 2 tsp of baking soda on top of the mix.

Slowly pour vinegar on the lemons and document your observations.

Fun Science Fact:

Baking soda (base) and vinegar and lemons (acid) create a chemical reaction by creating carbon dioxide (CO2) when mixed.

TRICKY TASTEBUDS

EDIBLE METER: Edible

TOOLS: cotton ball, (optional: Tricky taste buds worksheet)

VOCABULARY: 5 senses

QUESTION: How does your sense of smell affect your sense of taste?

INGREDIENTS

- Apple
- Vanilla extract

PROCEDURE

Document your hypothesis about this experiment.

Look at your apple slice and then eat it.

Close your eyes and smell the vanilla on a cotton swab and eat another apple slice.

Document your experience

Have extra fun with this activity with different food types and varieties.

Fun Science Fact:

Did you ever notice that when you are sick and can't smell, it directly affects your taste as well? That's because most of what we taste is directly related to what we smell.

Tricky Tastebuds

GROWING GUMMIES

EDIBLE METER: Taste safe

TOOLS: small bowls, sticky notes/labels, (optional: Growing gummies worksheet)

VOCABULARY: osmosis

QUESTION: What happens when gelatin comes into contact with water, salt water, sugar water, baking soda, vinegar and milk?

INGREDIENTS

- Gummy bears
- Water
- Salt
- Sugar
- Baking soda
- Milk
- Vinegar

Fun Science Fact:

Osmosis is the process where water wants to either come into or go out of the gummy bear to balance both sides of the membrane. When water enters the gummy bear, the sugar in the gummies absorbs liquid and forces it to "grow" rather than dissolve. Eventually, there is a balance, and osmosis stops.

PROCEDURE

Label each note as follows: water, salt water, sugar water, baking soda water, vinegar, and milk.

Caution: Hot! Pour ½ cup of hot water and dissolve with 1 tbsp salt into a labeled bowl. Let cool.

Caution: Hot! Pour ½ cup of warm water and dissolve with 1 tbsp sugar into labeled bowl. Let cool.

Caution: Hot! Pour ½ cup of warm water and dissolve with 1 tbsp baking soda into the labeled bowl. Let cool.

Pour ½ cup of vinegar into the labeled bowl.

Pour ½ cup of milk into the labeled bowl.

Record hypothesis and predictions on the Growing Gummies worksheet.

Observe gummies for approximately 12 hours.

Remove gummies from the bowl and place them on the Growing Gummies Worksheet in each respective box. Place an original gummy bear next to each gummy bear used for the experiment.. Observe and document the differences.

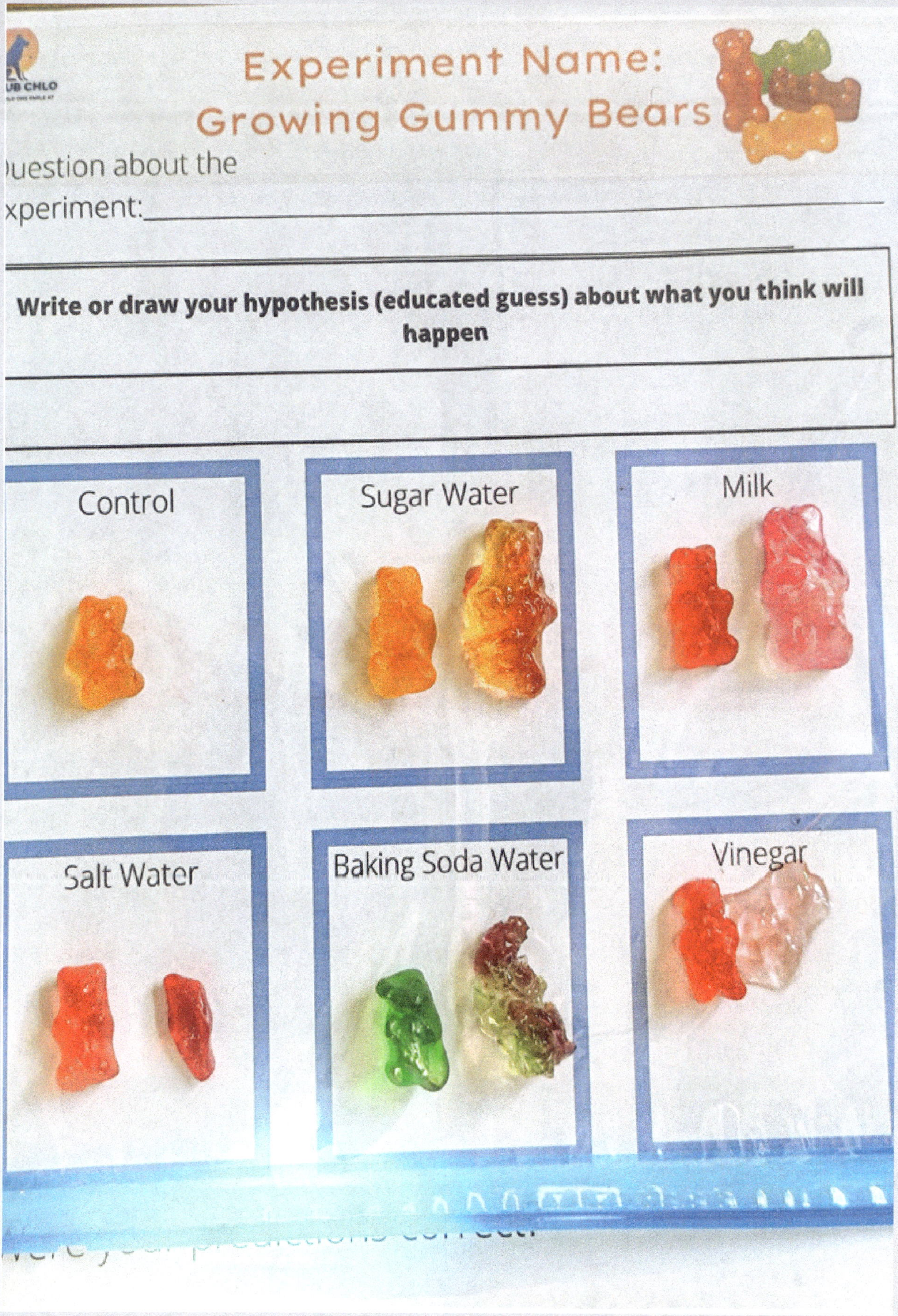

Experiment Name:
Growing Gummy Bears
uestion about the
xperiment:
Write or draw your hypothesis (educated guess) about what you think will happen
Control
Sugar Water
Milk
Salt Water
Baking Soda Water
Vinegar

TECHNOLOGY

Technology can be used to engage in digital activities such as computers, cell phones, robotics, coding, reverse engineering an electronic device and a host of other gadgets. But, did you know that science involving technology may not have to use electronics at all? A tool that is created to solve everyday problems is also considered using technology. Take a look around your kitchen and notice your sink, plates, spoons, measuring cups, etc, they were all tools created to solve a problem.

In this section, you will learn how to make ice and gelatin glow in dark. You will learn how to use a simple plastic spoon to conduct static electricity, write invisible ink messages and more.

GLOW IN THE DARK ICE

EDIBLE METER: Taste safe

TOOLS: Blacklight, ice tray

VOCABULARY: fluorescence, quinine, Ultraviolet light (UV)

QUESTION: How can you use tonic water to make ice glow in the dark?

INGREDIENTS

- Tonic water with quinine

PROCEDURE

Fill the ice cube tray with tonic water and freeze.

Place frozen cubes on a plate and enter a dark room

Turn on Blacklight and observe the ice.

Note:

low powered UV light is safe to use. As with all lights, please avoid shining directly in eyes.

Fun Science Fact:

Special light from the Blacklight is called ultraviolet light. Humans cannot see this special light. Quinine is a fluorescent substance that can absorb the invisible UV light and reflect it in a way that humans can see.

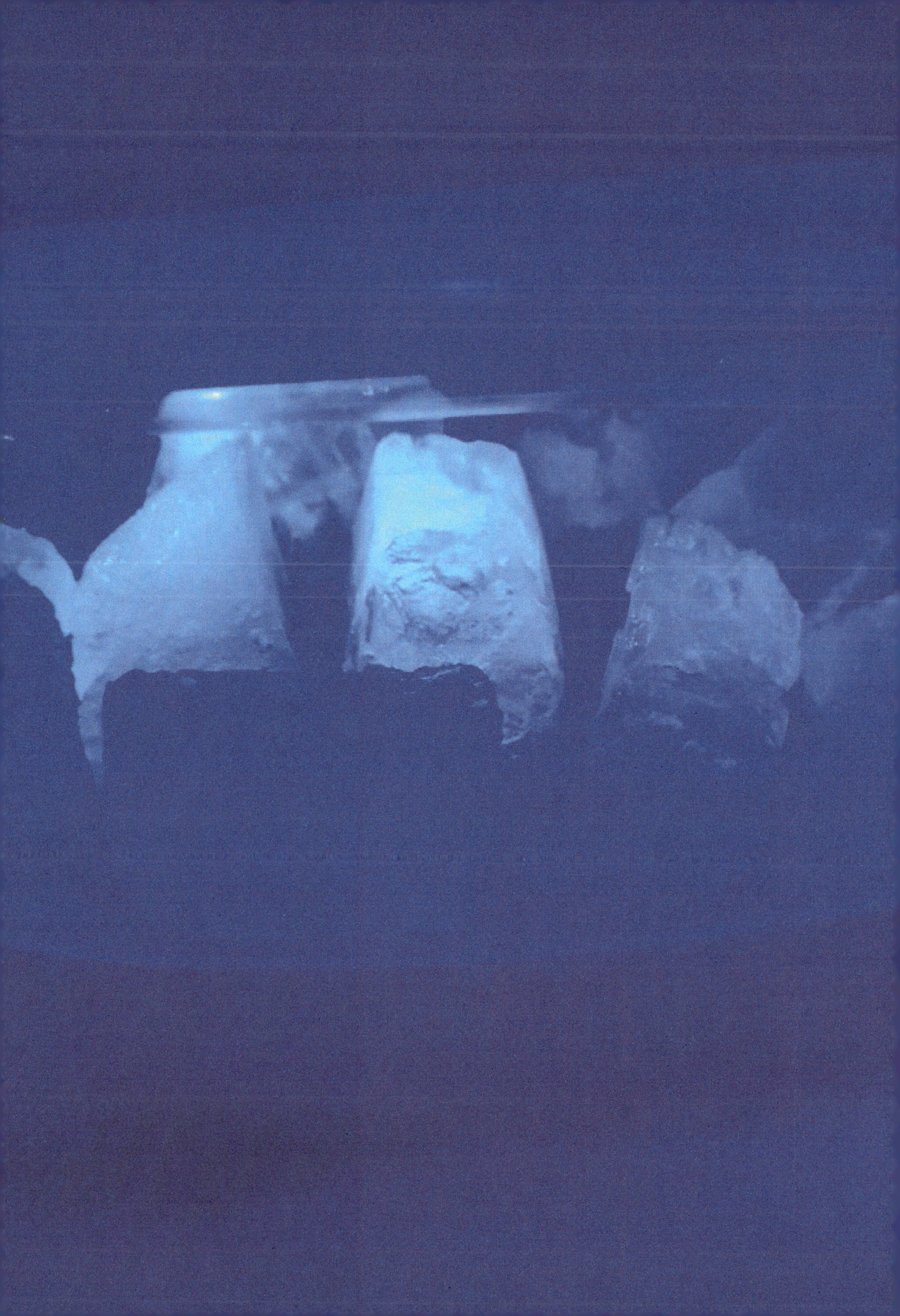

GLOW IN THE DARK GELATIN

EDIBLE METER: Taste safe

TOOLS: Blacklight, small pot, medium-sized bowl

VOCABULARY: fluorescence, quinine, Ultraviolet light (UV)

QUESTION: How can you use tonic water to make jello glow in the dark?

INGREDIENTS

- 8 oz Tonic water with quinine
- 1 packet of gelatin dessert powder (lemon or lime)
- Water

Note:

low powered UV light is safe to use. As with all lights, please avoid shining directly in eyes.

PROCEDURE

Caution: Hot! Boil 8 oz tonic water in a small pot.

Caution: Hot! Mix tonic water with a packet of gelatin dessert powder in a medium-sized bowl until dissolved.

Add 8 oz of cold water to the bowl and mix.

Refrigerate for 4 hours.

Enter a dark room, turn on Blacklight and observe the gelatin.

Fun Science Fact:

Special light from the Blacklight is called ultraviolet light. Humans cannot see this special light. Quinine is a fluorescent substance that can absorb the invisible UV light and reflect it in a way that humans can see.

FASCINATING SPOON STATIC ELECTRICITY

EDIBLE METER: Taste safe

TOOLS: Plastic spoon, dish cloth, (optional: Fascinating Spoon Static Electricity worksheet)

VOCABULARY: static electricity

QUESTION: Can salt and pepper be separated after being mixed together?

INGREDIENTS

- 1 tsp salt
- 1 tsp pepper

PROCEDURE

Sprinkle salt onto a solid colored paper. Then pour pepper onto pile of salt.

Rub plastic spoon on dish cloth for at least 10 seconds.

Hover the spoon over the salt and pepper pile and document your observations.

FUN!

Fun Tip:

You can also try this experiment using a blown up balloon instead of a plastic spoon.

Fun Science Fact:

Static electricity is when an electrical charge builds up on top of a static (unmoving) object over time. These electrical charges can be positive, negative or neutral. Rubbing the spoon gives it a negative charge; so, when the spoon comes into contact with particles of a different charge, they will be attracted to each other. Since pepper is lighter in weight than salt, you will notice more pepper particles than salt on the spoon.

INVISIBLE INK SECRET MESSAGE

EDIBLE METER: Edible

TOOLS: paintbrush/q-tip, white paper, (optional: Invisible ink secret message worksheet), stove/candle/lamp, small beaker/cup

VOCABULARY: citric acid, oxidation

QUESTION: How to use lemons to create a secret message?

INGREDIENTS

- Lemon

Note:

You can also hold the paper up to a lightbulb and wait for the message to appear.

PROCEDURE

Squeeze juice from the lemon into a small beaker.

Dip the q-tip into the juice and begin writing your secret message onto the paper. Set aside to dry.

Caution: Grown up needed to decode message: Hover the paper over an open flame such as candle or stovetop and message will appear.

Fun Science Fact:

Lemons contain a colorless acid called citric acid. So, when you write with it, it appears invisible. But, when you add heat to the acid some of the acid oxidizes (escapes into the air) and turns brown. This is the same oxidation process that happens when apples turn brown after being exposed to oxygen.

I love
you
Karter
50ml

ENGINEERING

Engineers use math and science everyday to design and solve their problems. There are 4 main types of engineers; chemical, civil, mechanical and electrical.

In this chaper, we will explore 3 types of engineering to be used in your kitchen lab. Chemical engineers can use their skills in the kitchen to help create and preserve foods. You will use chemical engineering to conduct a reversible chocolate science experiment,create your own ice cream, explore what happens when you microwave a marshmallow and even discover how to create your own snotty slime.

Mechnical engineers design machines like your refrigerator and microwave. You will become a mechanical engineer to build a blood pumping heart model, human DNA model, leak-proof bag and explore the design of an egg.

Civil Engineers design roads, bridges, and buildings that are safe for use. You will use civil engineering with grapes to build a sound stable structure.

EDIBLE DNA

EDIBLE METER: Edible

TOOLS: toothpicks, (Optional: DNA experiment worksheet)

VOCABULARY: DNA, chromosomes

QUESTION: What is the purpose of human DNA?

INGREDIENTS

- Strawberry licorice
- Gumdrops

Note:

for older kids, you could demonstrate the exact replica of a DNA strand while explaining what happens when pairs are duplicated.

PROCEDURE

Use 2 gumdrop colors and a toothpick to connect to the top of two licorice pieces. (Note: grapes, marshmallows, or any other small foods that come in multiple colors can be used instead.)

Continue pairing 2 gumdrop colors on individual toothpicks and placing them on licorice strands.

Use the DNA worksheet to color your DNA strand and use it as a guide for your DNA model. Or, use a DNA worksheet after creating a model to replicate the one you built.

Fun Science Fact:

DNA is an easy way of saying deoxyribonucleic acid and it is in the cells of every living thing called chromosomes. DNA carries all the important information about how a living thing will look and act.

GRAPE ENGINEERING

EDIBLE METER: Edible

TOOLS: toothpicks, (optional: Grape engineering worksheet)

VOCABULARY: engineering

QUESTION: How are buildings, bridges and houses constructed?

INGREDIENTS

- Grapes,

PROCEDURE

Draw a structure made from grapes and toothpicks and then see if you can build it in real life.

(Structure Ideas: build a bridge, abstract sculptures, simple shapes or research a famous structure and see if you can duplicate it.)

Fun Science Fact:

Did you know engineers use a similar process to build houses, tall buildings, towers, and even bridges? You are now an engineer too!

LEAK-PROOF BAG

EDIBLE METER: Non-Toxic

TOOLS: 3 sharpened pencils (or more), gallon-sized freezer bag (optional: Leak-proof Bag worksheet)

VOCABULARY: polymers

QUESTION: Why wont the zip top bag break when sharp pencils are added?

INGREDIENTS

- Water

PROCEDURE

Write your hypothesis about what you think will happen when the pencils are poked through the freezer bag filled with water.

Fill the freezer bag half full with water and begin sticking pencils straight through the bag. Document what happens.

Fun Science Fact:

Polymers are long chains of molecules. The plastic freezer bag is made of polymers that spread out and make a seal around the pencil preventing the bag from leaking when they are inserted.

WOW

BLOOD PUMPING HEART MODEL

EDIBLE METER: Edible (only the water)

TOOLS: water bottle/jars, 1 balloon, 2 plastic straws, rubber bands/tape, scissors, tray, (Optional: Blood Pumping Heart Model worksheet)

VOCABULARY: atrium, ventricles

QUESTION: How does the human heart work?

INGREDIENTS

- Water
- Red food coloring

Fun Science Fact:

The human heart is a pump: the right chamber of the heart receives blood from the body and pumps it back out to the lungs (atrium). The left chamber of the heart takes the blood from the lungs and pumps it out to the body (ventricles).

PROCEDURE

Fill jar ½ full with water and mix in 3-4 drops of red food coloring.

Cut the neck off the balloon and stretch over jar. Secure balloon with a rubber band.

Take the "neck" of the balloon that was cut and use a rubber band to secure the tip of a straw.

Cut two tiny holes in balloon stretched over jar and slide straws inside. Note: Make sure the holes are smaller than the straws so a tight seal is formed.

Place jar on top of the tray and gently begin "pumping" the heart by pressing the center of the balloon.

Document and observe reactions.

IMPOSSIBLE EGG

EDIBLE METER: Non-toxic (but not edible)

TOOLS: your hand, (optional: Impossible egg worksheet)

VOCABULARY: osmosis

QUESTION: What makes an egg so strong?

INGREDIENTS

- An egg

PROCEDURE

Place the egg in the palm of your hand and wrap your fingers around it and squeeze. Document your observation.

Fun Science Fact:

What makes eggs so strong? The secret is in the shape of the eggs. Eggs are shaped similarly to a 3-dimensional arch, which is one of the strongest building forms. The egg is strongest at the top and the bottom which is why it doesn't break when adding an even amount of pressure on both ends. This is why hens are able to sit on eggs and not crack them. However, eggs break easily when uneven pressure is applied like tapping it on its sides.

WOW!!
born scientist

REVERSIBLE CHOCOLATE SCIENCE

EDIBLE METER: Edible

TOOLS: Baking mold/wax paper

VOCABULARY: states of matter, reversible change

QUESTION: How can we change chocolate from a solid to a liquid?

INGREDIENTS

- Chocolate chips (optional: sprinkles)

PROCEDURE

Caution: Hot! Microwave 1 tbsp of chocolate chips in 30 second intervals

Pour chocolate into molds and refrigerate for at least 10 minutes.

(Note: if using sprinkles add to the mold first, then pour chocolate on top.)

Document how the states of matter changed? How could you reverse the change?

Fun Science Fact:

The chocolate chips were changed from a solid to a liquid. You can change the liquid back into a solid, that is called reversible change. An irreversible change is trying to change a cake back to all the raw ingredients. That's impossible.

ICE CREAM IN A BAG

EDIBLE METER: Edible

TOOLS: quart-sized freezer bags, gallon-sized freezer bags, (optional: ice cream in a bag worksheet)

VOCABULARY: freezing point, states of matter

QUESTION: What happens when you mix salt and ice?

INGREDIENTS

- ½ cup half and half (or coconut milk)
- 1/3 cup rock salt
- ½ tsp vanilla extract
- 1 tbsp sugar
- 3 cups of ice

PROCEDURE

Place ice and salt into a gallon-sized freezer bag and set aside.

In quart-sized freezer bag mix half and half, vanilla and sugar, and seal.

Place a quart-sized bag inside a gallon-sized bag and shake. Use gloves or a kitchen towel as the bag gets cold. (Pro tip: turn on some dance music and shake for an entire song.)

Document your observations

Fun Science Fact:

Salt and ice mixture create a chemical reaction that changes liquids into solids and forces the contents to freeze. Salt lowers the freezing point of water and freezes all contents that come into contact with it.

CAKE CHEMISTRY

EDIBLE METER: Edible

TOOLS: sticky notes, pen/marker, (Optional: "Kitchen Cake Chemistry" worksheet)

VOCABULARY: carbon dioxide, fat

QUESTION: What happens if oil, egg or baking powder are removed from a recipe?

INGREDIENTS

- 6 tsp flour
- 3 tbsp sugar
- 1 dash of salt
- 2 dashes baking powder
- 2 tbsp milk
- 2 tbsp cooking oil
- ¼ tsp vanilla
- 1/3 egg (beat egg and divide)

Note

Measure and mix ingredients four times for each individual mini cake

PROCEDURE

Label each small dish with a sticky note to keep track of ingredients (all ingredients, no oil, no baking powder, no egg).

Mix dry ingredients then add wet ingredients.

Bake at 350 for 20 minutes.

Document what you think will happen after each cake is removed from the oven.

Repeat steps for remaining cakes. View labels to help omit specific ingredients.

Caution: Hot! Remove cakes from the oven and let them cool. How does each cake look? How does each cake taste?

Fun Science Fact:

Baking powder releases carbon dioxide and helps the cake to rise. Oil contains fat which helps the cake to be soft and moist after being baked. Eggs add color, flavor,, and texture to baked goods. Eggs also contain fat which helps baked goods remain tender.

Everything
No oil
No egg
No Baking
Powder

MICROWAVED MARSHMALLOW

EDIBLE METER: Edible

TOOLS: microwave, medium-sized bowls, or large plate

VOCABULARY: molecules, volume

QUESTION: Why are marshmallows so fluffy?

INGREDIENTS

- 3 jumbo marshmallows

PROCEDURE

Microwave each marshmallow with the following times: 15 seconds, 30 seconds, 60 seconds (Note: use and edible marker to write the amound of time each marshmallow was microwaved)

Document on the worksheet your hypothesis about what you think will happen when each marshmallow is microwaved.

Caution: Hot! Microwave the first marshmallow for 15 seconds and document your observation.

Repeat the process for the remaining marshmallows.

Fun Science Fact:

Marshmallows are so fluffy because of all their air bubbles, sugar, and water. When marshmallows are heated (temperature increases), their water molecules and air bubbles expand very quickly (volume increases). This softens the sugar and makes the marshmallow very stretchy.

ART

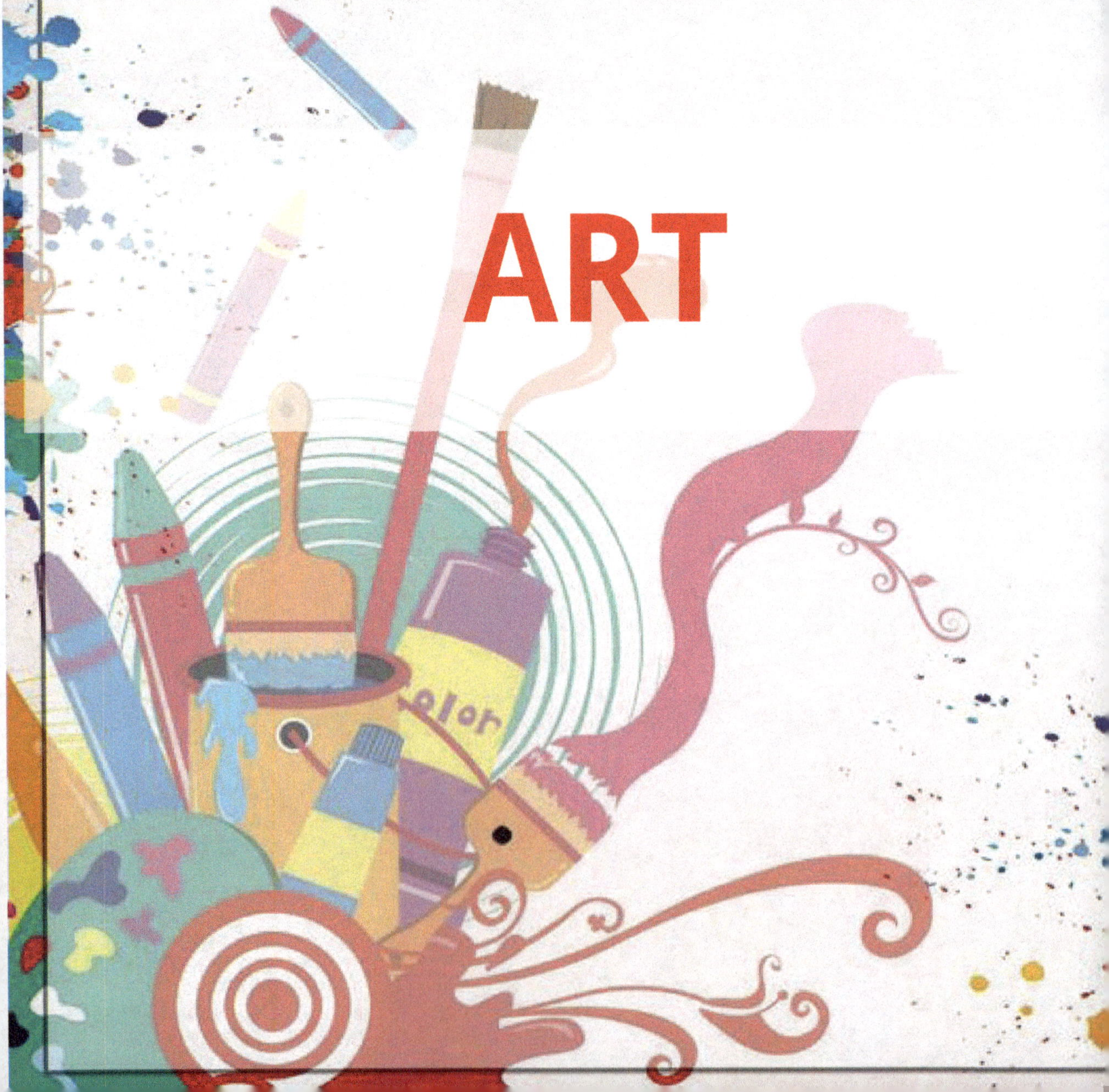

Adding color to projects helps to encourage creative and adds overall beauty to the experiments.

Keep your lab coats on and take our your artist's painter's palette as you incorporate art into these science projects.

In this section, you will explore the colorful rainbow of colored candies, flying colors in a jar and pasta slime.

Have fun creating your own colorful milk, sugar sand, and lava lamp made out of ice cubes.

Paint your own fizzy earth and discover how to paint on ice.

Can't wait to see all your creative experiments

COLORED CANDY CHROMATOGRAPHY

EDIBLE METER: Edible

TOOLS: ice tray, jar, (optional: Lava Lamp observation worksheet)

VOCABULARY: density

QUESTION: Why wont oil and water mix?

INGREDIENTS

- Water
- Food coloring
- Vegetable oil

PROCEDURE

Fill an ice tray with water and add 3-4 drops of food coloring in each space and mix freeze.

Fill jar ¾ full with vegetable oil, drop ice cubes inside and stir

Document and observe the reaction

Fun Science Fact:

Why won't oil and water mix? Oil is less dense than water and also not water-based so the two substances will never mix instead they will slowly separate from each other

ICE CUBE LAVA LAMP

EDIBLE METER: Taste safe

TOOLS: large glass or ceramic tray (optional: colored candy chromatography)

VOCABULARY: chromatography

INGREDIENTS

- Multi-colored chocolate candies
- Warm water

PROCEDURE

Arrange candies around the edge of the tray/plate.

Fill a glass with warm water.

Pour water into the center of the plate.

Document your observations.

Fun Science Fact:

chromatography is the chemical process in which components are separated. As the warm water touches the candies the sugar begins to dissolve and the colors spread as it mixes with water.

Fun Tip:

if you are in a place with multiple children such as a classroom or kids party, break kids into groups and set up multiple experiments. Have the junior scientists set their timers and record data. Will each group have the similar data or dramatic differences? Let the fun begin!

1 PT
3/4
1/2

FLYING COLORS IN THE JAR

EDIBLE METER: Taste safe

TOOLS: Jar, (optional: Flying colors in a jar worksheet)

VOCABULARY: Density

QUESTION: What happens when you mix oil and water?

INGREDIENTS

- 1 cup water
- 4 Tbsp cooking oil
- Food coloring

PROCEDURE

Fill jar with ½ cup of water

Pour cooking oil into a jar

Add 4-5 drops of food coloring for each color.

Document observations.

Fun Science Fact:

Oil and water will not mix due to different densities. Water is higher in density and oil is lower in density. Lower density materials will float. Food coloring is water-based and higher in density than oil, so it will drop through the oil and mix with the water.

PASTA SLIME

EDIBLE METER: Edible

TOOLS: pots, small cups/bowls, large dish/tray

VOCABULARY: absorption

QUESTION: How does spaghetti turn from being a hard solid to soft and squishy?

INGREDIENTS

- Spaghetti
- Food coloring

PROCEDURE

Divide spaghetti into 3 parts. (one part for each color)

Add 3-4 cups of water and 5-10 drops of food coloring into each pot and bring to a boil. (note: flavored juice mix packets can work as well.)

Add spaghetti and cook for about 10-15 minutes. The spaghetti will turn colors

Drain, rinse and add a few drops of cooking oil.

Fun Science Fact:

As the spaghetti cooks, it absorbs all liquid in the pot. When the liquid in the pot is colored, the spaghetti will turn colors as well.

FIZZY PAINT EARTH CRAFT

EDIBLE METER: Non-Toxic

TOOLS: paintbrushes, pipettes, small cups, pencil, white crayon (optional: fizzy paint earth craft worksheet)

VOCABULARY: base, acid, chemical reaction, carbon dioxide (CO2)

QUESTION: Can a painting be created using baking soda and food coloring? What happens when vinegar is added?

INGREDIENTS

- Baking soda
- Vinegar
- Black construction paper
- Food coloring

PROCEDURE

Mix food coloring with 1 tbsp baking soda (for every color).

Use a bowl to outline a circle and trace with a pencil on cardstock.

Use paintbrushes to pack a thick layer of paint onto cardstock.

Draw stars around the earth with a white crayon and let the paint dry.

Fill the pipette with vinegar and slowly release it onto the earth painting. Observe the reaction. What do you see? What do you hear? What do you feel?

Fun Science Fact:

Baking soda (base) and vinegar (acid) create a chemical reaction by creating carbon dioxide (CO2) when mixed.

The Earth is the 3rd planet from the Sun and was formed over 4.5 billion years ago.

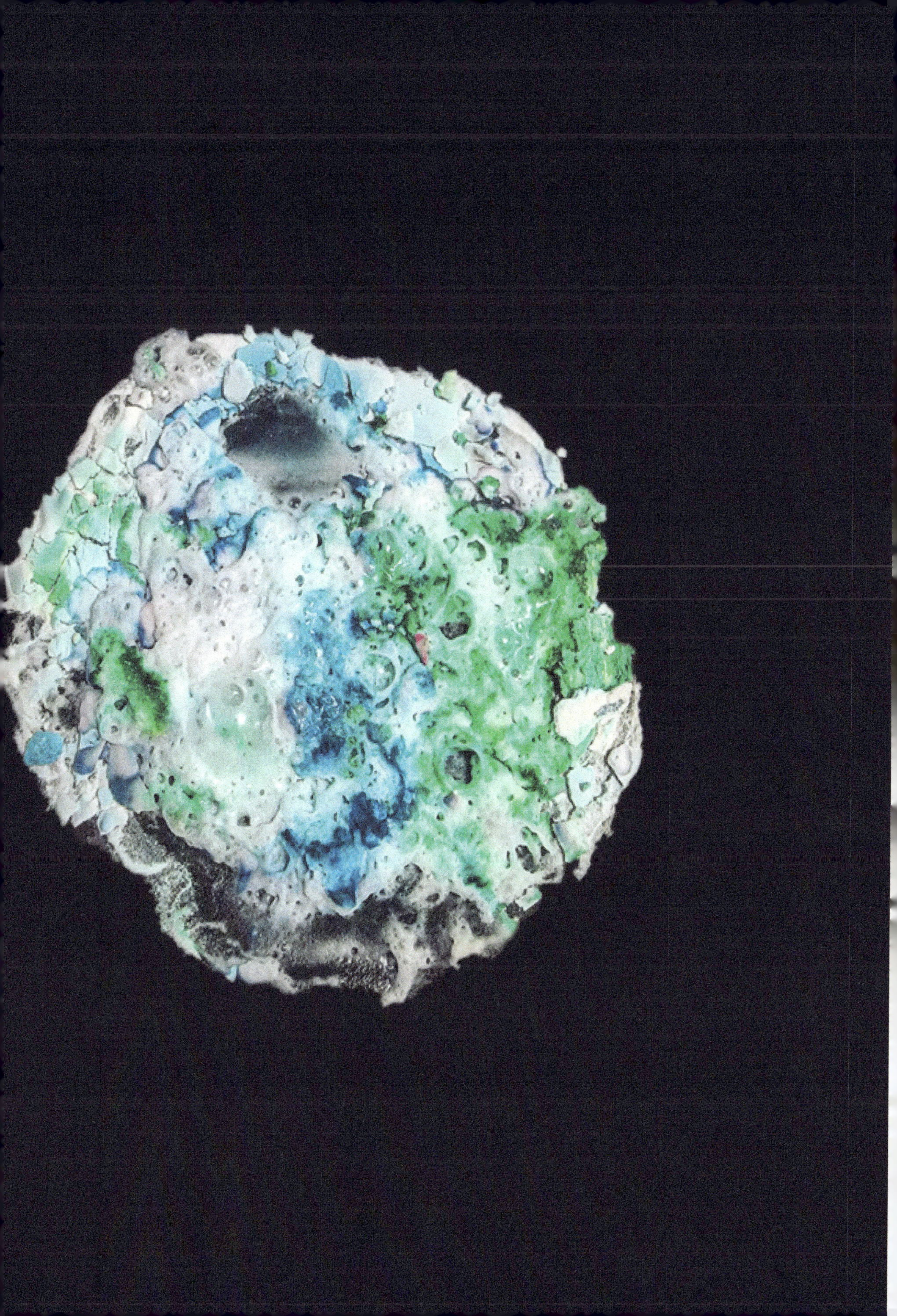

ICE AND SALT EXPERIMENT

EDIBLE METER: Taste-Safe

TOOLS: paintbrush, bowl, (Optional: "Ice and salt" observation worksheet)

VOCABULARY: freezing point

QUESTION: What happens when salt is mixed with ice?

INGREDIENTS

- 2 tsp ice block
- Food coloring
- Course/rock salt

PROCEDURE

Use a plastic bowl to freeze an ice block at least 6 hours.

Mix 1 tbsp water with 3-4 drops of food coloring (note: separate bowl for each color).

Pour salt onto the ice block and use ice then record your observations. What do you see? What do you hear? What do you feel?

Use a paintbrush to color your masterpiece and document the process.

Fun Science Fact:

Why does salt melt ice? The freezing point of water is 0°c. The molecules in salt move much faster than the molecules in water, so when you add salt to ice, it lowers the freezing point even more and prevents ice from forming. This is why salt is put on icy roads after a major snowstorm.

VAN GOGH MILK (MILK DANCE)

EDIBLE METER: Taste Safe

TOOLS: shallow dish, cotton swab (q-tip or toothpick), Van Gogh Milk worksheet

VOCABULARY: Molecules

QUESTION: Why doesn't food coloring sink?

INGREDIENTS

- Half and half

Note:

adding dish soap makes the experiment no longer taste safe.

PROCEDURE

Pour 1 cup of milk into a shallow dish (make sure dish is fully covered.)

Add various food coloring drops in the center of the milk

Document what you think will happen if a cotton swab touches the colors on the milk.

Tap the cotton swab onto the food coloring. Document what happens.

Document what you think will happen if a drop of dish soap is added to the cotton swab.

Add a drop of dish soap to the cotton swab and observe what happens to the colors. Play around with the measurements of dish soap and colors for various effects, color theories, and even shapes.

Fun Science Fact:

Why doesn't the food coloring sink? What makes the colors dance across the milk? Milk is mostly water and contains proteins and fats. When the dish soap touches the milk, the proteins and fat start to break up. Then the soap molecules chase the fat molecules in the milk. This process is usually invisible to the human eye, but the food coloring helps you to see the entire process.

SUGAR SAND

EDIBLE METER: Edible

TOOLS: baking pan, quart-sized freezer bag

VOCABULARY: carbohydrates

QUESTION: What is the purpose of sugar?

INGREDIENTS

- ½ cup sugar
- Food coloring

PROCEDURE

Mix sugar and 5-8 drops of food coloring in the freezer bag and shake to mix well. (note: the process will have to be repeated for each color)

Let it sit for approximately 10 minutes to dry.

Fun Science Fact:

Sugars are sweet and taste great when mixed with certain foods. Sugars are carbohydrates that provides a great source of energy to humans; it travels through your blood stream and cells store it as energy for later. But too much sugar can cause tons of health problems. So, try to find a balanced way to consume sugar.

WALKING WATER

EDIBLE METER: Edible

TOOLS: 7 small cups, folded paper towels (optional: Walking Water worksheet)

VOCABULARY: capillary action

QUESTION: How are some colors created?

INGREDIENTS

- Food coloring
- Water

Note:

speed of experiment depends on how much water you have in your cups, and also how long and thick your paper towels are.

PROCEDURE

Fill 4 small cups full of water

Add 3 drops of blue food coloring to one cup, 3 drops of yellow to another cup, and 3 drops of red to two additional cups. Fill the other 3 cups with water and line them up as follows:

Red, empty cup, yellow, empty cup, blue, empty cup, red

Fold paper towels - place one side in the colored cups and the other in the empty cup, connecting them both.

Wait a few hours and document your observation.

Fun Science Fact:

Capillary action is what helps the colors to "walk" up the paper towel. This is the same process that trees and plants use to absorb water.

MATHEMATICS

Food scientists need math to help them make the perfect recipes and measure exact ingredients. Imagine trying to make a pizza pie with only a little bit of cheese. You will just have a large amount of dough - and that's no fun!

In this section, you will use your math skills to create many fun projects such as: Fizzy lemonade, Color changing lemonade, Hot chocolate playdough and edible water beads.

Calculate everything from how many days are in each Moon Phase to how many seconds it takes for a marshmallow to explode.

Learn the temperature needed to melt a marshmallow snowman and which ingredient can stopo an apple slice from turning brown.

There's so much fun packed in this section. Get your lab coats, pencils and worksheets and get ready to have a blast.

MARSHMALLOW PLAYDOUGH

EDIBLE METER: Taste safe

TOOLS: microwave, medium-sized bowl

VOCABULARY: molecules, volume

QUESTION: Can play dough be created using marshmallows?

INGREDIENTS

- 3 tbsp corn starch
- 6 large marshmallows
- 1 tsp coconut oil
- Food coloring

PROCEDURE

Combine all ingredients into a bowl and microwave for 30 seconds (Note: You will need a separate bowl for each color)

Caution: Hot! Use a spoon to mix all ingredients

Continue kneading the dough by hand

Fun Science Fact:

Marshmallows are so fluffy because of all their air bubbles, sugar, and water. When marshmallows are heated (temperature increases), their water molecules and air bubbles expand very quickly (volume increases). This softens the sugar and makes the marshmallow very stretchy.

ONE MINUTE CAKE IN A CUP

EDIBLE METER: Edible

TOOLS: mug, spoon, microwave (optional: one minute cup cake worksheet)

VOCABULARY: Carbon dioxide

QUESTION: What is the fastest time a cake can be baked?

INGREDIENTS

- ¼ cup flour
- 2 tbsp cocoa powder
- 2 tbsp sugar
- ¼ tsp baking powder
- pinch of salt
- 1/3 cup milk
- 2 tbsp oil/butter
- splash of vanilla (optional: 1 egg)

Safety tip:

If using an egg, microwave an additional 30 seconds.

PROCEDURE

Mix all ingredients together in a mug and microwave on high for 1 minute.

Let it sit for approximately 2 minutes before serving.

Note:

you might have to microwave an additional 20 seconds or more, depending on the size of your mug and according to taste.

Fun Science Fact:

Baking powder releases carbon dioxide and helps the cake to rise. Oil contains fat which helps the cake to be soft and moist after being baked. Eggs add color, flavor, and texture to baked goods. Eggs also contain fat which helps baked goods remain tender.

APPLE REACTIONS

EDIBLE METER: Taste safe

TOOLS: 5 small bowls/cups, (optional: "Apple Reactions" worksheet)

VOCABULARY: ascorbic acid

QUESTION: Why do apple slices turn brown? What can stop them?

INGREDIENTS

- 1 egg,
- 1 tbsp white vinegar
- 1 tbsp baking soda
- 1 tbsp milk
- 1 tbsp lemon juice

PROCEDURE

Pour each ingredient into an individual bowl and label.

Place each apple slice inside the individual bowl. **(Note: one apple slice is the control subject and will be placed inside an empty bowl.)**

Let apples sit for approximately one hour and use the "Apple Reactions" worksheet to document your observations

Fun Science Fact:

Lemon juice slows down the browning from occurring in apples due to the acid (ascorbic acid) in the liquid. Oxygen will interact with the acid in the lemon juice first before interacting with the enzymes in the apple. After all the acid has been used, then browning will occur as the oxygen begins interacting with the apples in the enzyme.

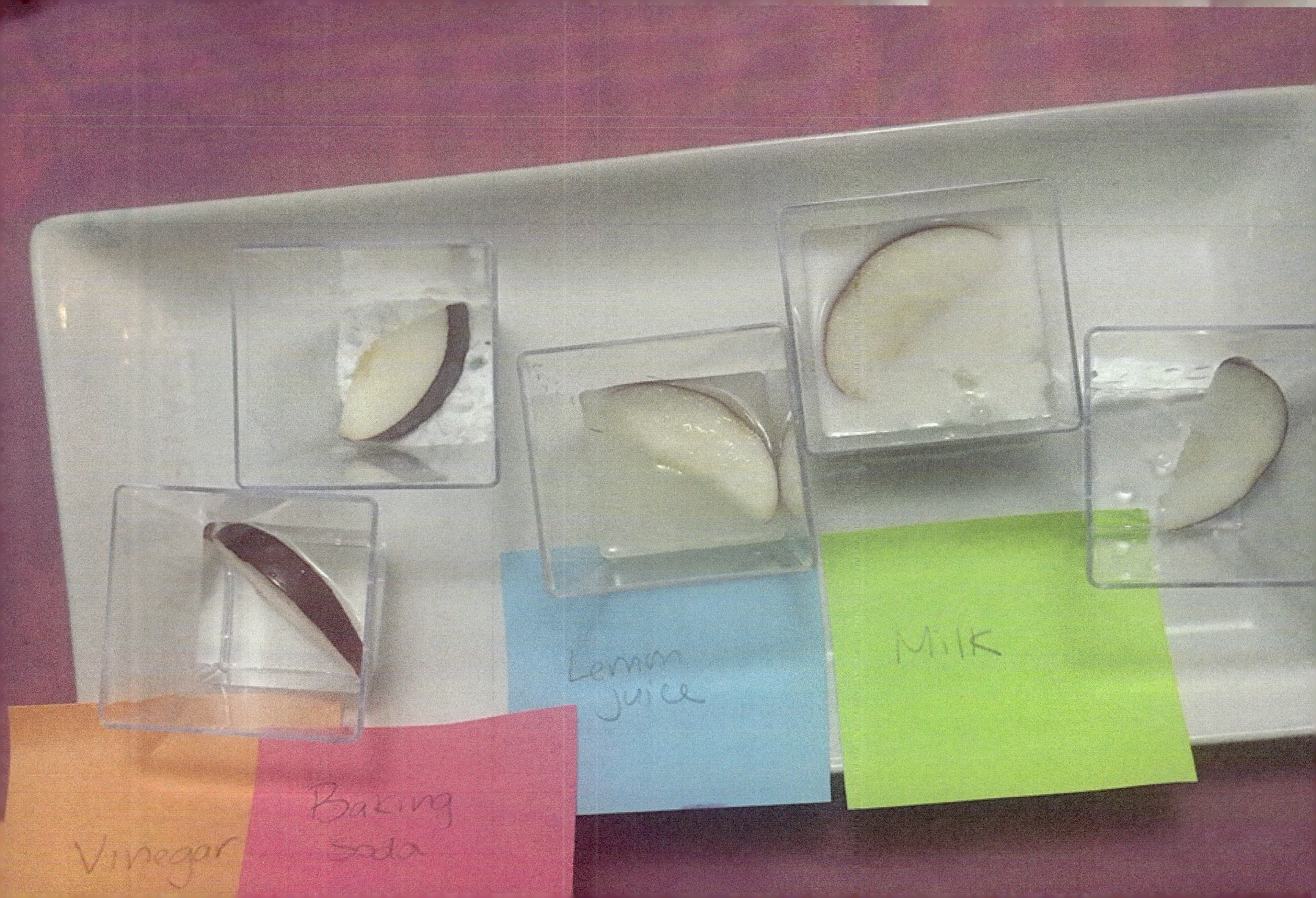
Vinegar
Baking Soda
Lemon juice
Milk

FIZZY LEMONADE

EDIBLE METER: Taste safe

TOOLS: spoon, glass, (optional: Fizzy Lemonade Worksheet)

VOCABULARY: base, acid, carbon dioxide

QUESTION: What happens when baking soda and lemonade are mixed?

INGREDIENTS

- 1 cup pre-made lemonade (or 1-2 lemons juiced)
- 1 tbsp baking soda
- Cold water
- 1-2 tsp sugar

PROCEDURE

Pour 1 cup of lemonade into a glass.

Add 1 tsp baking soda and mix. Add sugar to taste

Document your observations and conclusions.

Fun Science Fact:

Baking soda (base) and lemonade (acid) create a chemical reaction - this is the same chemical reaction you get when mixing baking soda and vinegar - carbon dioxide (CO2) is still created when mixed.

WOW!

COLOR CHANGING LEMONADE

EDIBLE METER: Edible

TOOLS: spoon, 4 glasses, small pot

VOCABULARY: anthocyanin, pH balance

QUESTION: Why does this lemonade change colors?

INGREDIENTS

- 4 cups lemonade
- 1 butterfly tea bag or 1 heaping tbsp of butterfly tea flowers
- Ice

PROCEDURE

Caution: Hot! Place tea inside boiling pot of water and let steep for 4 minutes.

Add ice cubes to glasses and pour ¾ cup full of lemonade

Slowly pour tea into the glasses and observe the reaction.

Fun Science Fact:

Butterfly pea flowers are primarily found in Thailand. They are special because they contain a blue pigment called anthocyanin which helps the tea to change colors as the pH (acidity) changes. The tea can be bright blue or even purple depending on the pH balance of the liquid it is mixed in.

Pro Tip:

have fun changing the pH balance of the tea by adding baking soda or other ingredients with various PH balances

ORANGE BUOYANCY

EDIBLE METER: Edible

TOOLS: tall glass, (Optional: "Orange Buoyancy" worksheet)

VOCABULARY: buoyancy, density

QUESTION: Why do some objects sink and others float when placed in liquid?

INGREDIENTS

- 1-2 oranges
- Water

PROCEDURE

Fill tall jar ¾ full with water.

Write your hypothesis on your worksheet about what you think will happen when the unpeeled orange is dropped into the glass.

Place the unpeeled orange in the glass and document what happens.

Write your hypothesis on your worksheet about what you think will happen when the orange is peeled and dropped again into the glass?

Peel the orange and place it into the water; document what happens.

Fun Science Fact:

Buoyancy is whether something has the ability to sink or float in the liquid. Orange, with the peel on, floated because it had tiny pockets of air inside the peel that made it less dense than the water. But, when you removed the peel, it sank because it became denser or heavier than the water.

HOT CHOCOLATE PLAYDOUGH

EDIBLE METER: Taste-Safe

TOOLS: bowl, spoon

VOCABULARY: proteins

QUESTION: Can playdough be created from hot chocolate mix?

INGREDIENTS

- 1 packet of hot chocolate mix
- 2 cups of flour
- ¾ cup salt 1 cup baking soda
- ½ cup water
- 3 tbsp water
- ¼ cup oil

PROCEDURE

Mix all dry ingredients in a bowl. Then add oil.

Next, slowly add water to the bowl and mix. Then, knead with hands and form into a ball.

Place playdough ball onto a floured surface and continue to knead until mixture is smooth and brown.

Fun Science Fact:

Salt helps proteins stick together. Flour contains proteins. When water is added to flour, the proteins and the salt in the flour stick together. This gives you the same "dough-like" feeling just like when you are forming cookies. You might be able to smell the salt in your playdough just like when you buy playdough at the store.

MELTING SNOWMEN

EDIBLE METER: Edible

TOOLS: small cups, marker, black edible marker, orange edible marker, (optional: melting snowman observation worksheet, stopwatch)

VOCABULARY: changes of matter, molecules, kinetic energy

QUESTION: Which temperature will melt a marshmallow the fastest: hot, warm or cold?

INGREDIENTS

- Hot cocoa mix
- 3 large marshmallows
- 3 mini marshmallows
- Milk or water

PROCEDURE

Use edible markers to draw faces on your snowmen

Prepare one cup of hot cocoa and pour it into a cup labeled "hot."

Prepare another cup of cocoa and microwave milk for a shorter time period for warm liquid. Pour into cups labeled "warm" and "cold."

Add an ice cube to the "cold" cup.

Drop the mini and jumbo marshmallows into the hot cocoa cup and start the stopwatch. Use the "Melting Snowman Observation" worksheet to document how long it took each snowman to melt. Which size melted the fastest? Repeat process for remaining cups.

Fun Science Fact:

When something is hot in temperature the molecules inside it are moving faster, creating more kinetic energy. Melting is when a solid turns into a liquid. When the solid marshmallow comes in contact with the hot liquid, the molecules in the marshmallow begin to heat quickly and will turn into a liquid. The smaller the solid, the faster it will turn to liquid. When placed in colder liquid the molecules move more slowly and will remain as is.

Hot
Warm
Cold

GUMMY BEAR SLIME

EDIBLE METER: Taste safe

TOOLS: microwave-safe bowl

VOCABULARY: polymers

QUESTION: What happens when you heat up gummy bears in the microwave?

INGREDIENTS

- ½ cup gummy bears (each color selected)
- 2 tbsp cornstarch
- 1 tbsp powdered sugar

PROCEDURE

Melt and mix ½ cup of gummy bears in 15-second intervals in the microwave. (Note: separate bowl needed for each color)

Mix 2 tbsp cornstarch and 1 tbsp powdered sugar and knead. (Note: If you want your gummies to feel more like dough than slime, add additional cornstarch and powdered sugar.

Fun Science Fact:

The heating of gelatin creates longer polymers. When the polymers are mixed with cornstarch, the cornstarch serves as an activator and gives the gummies the slime-like consistency while the sugar changes the texture to feel more like dough.

PARTY CAKE FROSTING PLAYDOUGH

EDIBLE METER: Edible

TOOLS: medium-sized bowl, spoon, (optional: food coloring)

VOCABULARY: osmosis

QUESTION: Can playdough be created using sugar and cake frosting?

INGREDIENTS

- 1 cup vanilla frosting
- ½ cup cornstarch
- ½ cup confectioners sugar (icing sugar)

PROCEDURE

Mix cornstarch and sugar into the bowl and add 3-5 drops of food coloring

Add 1 cup of vanilla frosting to ingredients and mix well.

Adjust texture by adding sugar until desired consistency is reached and knead.

Fun Science Fact:

Here's a fun baking science trick: You can make your own frosting with a potato. Bake and mash the potato, then mix with powdered sugar and vanilla and enjoy!

EDIBLE WATER BEADS

EDIBLE METER: Edible

TOOLS: small pot, small bowls, bin/large tray

VOCABULARY: absorption

QUESTION: Why are water beads so squishy?

INGREDIENTS

- Quick-cook tapioca pearls
- Water
- Food coloring

PROCEDURE

Bring a medium-sized pot to boil, add tapioca pearls, stir, cover pot and cook for 4-7 minutes.

Drain pearls and rinse with cold water

Divide pearls into cups and add several drops of food coloring and mix. Let stand for 15 minutes then rinse with cold water. (Pro tip: if too sticky add water to the bin

Fun Science Fact:

When Tapioca pearls are cooked, they swell in size as they absorb water, become soft and squishy and resemble water beads. The difference is tapioca pearls are non-toxic if ingested, in fact, they are completely edible.

SNOTTY SLIME

EDIBLE METER: Edible

TOOLS: bowl

VOCABULARY: polymer

QUESTION: What is mucus?

INGREDIENTS

- 3 packs of unflavored gelatin
- Corn syrup
- Water
- Food coloring

PROCEDURE

Boil ½ cup of water, add 5 drops of green food coloring and let it sit for 5 minutes.

Slowly add gelatin to bowl and mix lightly with a fork.

Slowly add ½ cup of corn syrup to the mixture until desired consistency is reached.

Use a fork to pull up the snotty solution and gross everyone out.

Fun Science Fact:

The mixture of gelatin and water creates a polymer. When the polymer is mixed with corn syrup, the strands become long and string-like which resembles mucus.

Mucus protects your lungs and nose by trapping dust, dirt and pollen from entering your body. Mucus can be found in all living organisms.

WOW

DANCING RAISINS

EDIBLE METER: Edible

TOOLS: Water bottle/jars, dancing raisins worksheet, (Optional: food coloring, "Dancing Raisins" worksheet)

VOCABULARY: buoyancy

QUESTION: Do raisins sink or float when added to carbonated water?

INGREDIENTS

- Raisins
- Clear carbonated soda

PROCEDURE

Fill water bottle ¾ full with carbonated soda and drop a small handful of raisins into the bottle.

Document and observe the raisins' reactions.

Fun Science Fact:

Buoyancy is whether something has the ability to sink or float in the liquid. Raisins sank because they were heavier than the liquid, but the carbon dioxide in the soda created bubbles that caused the raisins to "dance" while rising to the top. Once reached to the top, the bubbles popped and the raisins fell back down again.

SINK OR SWIM EGGS

EDIBLE METER: Non-Toxic

TOOLS: 2 glasses, (Optional: "Sink or Swim Eggs" worksheet)

VOCABULARY: density

QUESTION: Do eggs sink or float when placed in water?

INGREDIENTS

- 2 eggs
- Water
- 3 tbsp salt

PROCEDURE

Fill one glass ¾ full with water.

Write your hypothesis on your worksheet about what you think will happen when the egg is dropped into the glass. Will it sink or float?

Place the egg in the glass and document what happens.

Write your hypothesis on your worksheet about what you think will happen when the egg is placed into the glass filled with saltwater?

Add salt to the water and carefully place the second egg into the water; document what happens.

Fun Science Fact:

Density is how much "stuff" is packed into an object making it heavier or lighter. An egg will sink in water because it is denser than water. But, when you add salt to the water, the liquid becomes denser than the egg and the egg will float.

MARSHMALLOW SLIME

EDIBLE METER: Taste safe

TOOLS: Microwave

VOCABULARY: molecules

QUESTION: What happens when you heat up a marshmallow in the microwave?

INGREDIENTS

- 2 Jumbo marshmallows
- 1 tsp cooking oil
- 2 tbsp cornstarch
- (optional: food coloring)

PROCEDURE

Caution: Hot! Microwave 2 jumbo marshmallows for 30 seconds.

Mix ½ tsp of oil into melted marshmallows.

Add 1 tbsp cornstarch at a time and knead until your desired consistency is reached. (note: you can replace cornstarch with powdered sugar if your young scientists want to eat their experiment in larger quantities.)

Fun Science Fact:

When the marshmallow is heated, the water molecules make it easier to stretch and play with. But, the more it is constantly stretched, it loses its water molecules and begins to harden.

CHOCOLATE COOKIE MOON PHASES

EDIBLE METER: Edible

TOOLS: butterknife, paper plate, (optional: man on the moon: moon phases worksheet)

VOCABULARY: moon phases

QUESTION: How many days does it take for the moon to cycle around the earth?

INGREDIENTS

- Chocolate cream-filled cookies

PROCEDURE

Slowly separate 8 chocolate cookies from the cream inside.
Use a butter knife to carve out each phase of the moon.

Fun Science Fact:

Every 29.5 days the moon completes a full cycle around the earth. Each phase is a different position of the moon's orbit. As the moon orbits the earth, the shadows from the sun cast onto the moon change our full visibility with each phase. Research each phase of the moon and mark on your calendar when the next full moon is expected to occur.

Experiment Name: Man on the Moon - Moon Phases

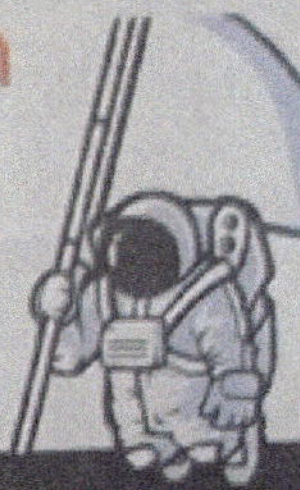

SOIL LAYERS

EDIBLE METER: Edible

TOOLS: dessert cup/bowl, zip-top bag, rolling pin

VOCABULARY: bedrock, topsoil, subsoil, microbes

QUESTION: Where does the dirt come from?

INGREDIENTS

- Chocolate cream-filled cookies
- Gummy worms
- Graham crackers
- Multi-colored chocolate candies
- Sprinkles

PROCEDURE

Bedrock: Fill the bottom of the dessert cup with brown chocolate-covered candies.
Subsoil: Crush a Graham cracker and layer it on top of the bedrock
Topsoil: Use a zip-top bag and rolling pin to crush dark chocolate cream-filled cookies.
Living organisms: Place gummy worms, sprinkles, and colored chocolate candies on top of topsoil.

Fun Science Fact:

Bedrock hard rock that is buried under thousands of layers of earth. The subsoil is hard dirt and rock that is also layers deep and sits on top of the bedrock. Plants with deep roots can often reach this part of the earth. Topsoil is not as dense as the other layers, in fact, it is only a few inches deep which makes it the thinnest and richest layer where nutrients are found and seeds are planted. Living organisms such as worms, bugs, and microbes break down the nutrients in the soil that help the plants grow.

GLOSSARY

Absorption - To soak something up.

Acid - is any substance that is sour in water solution, reacts with metals, bases and promotes chemical reactions.

Anthocyanin - Chemical that makes the color in plants red.

Atrium - Top chambers of the heart that squeezes blood into the heart.

Base - A substance that can neutralize acids.

Bedrock - (subsoil, topsoil, microbes)

Buoyancy - Buoyancy is whether something has the ability to sink or float in liquid.

Capillary Action - A process where liquid moves up something solid.

Carbohydrates - A compound made of carbon, hydrogen and oxygen. It is formed by plants and used in the body to produce and store energy. Sugars, starches.

Carbon Dioxide - A molecule that contains two oxygen atoms and one carbon atom that can come in many forms: the air we exhale, gas in the air, and other forms.

Chemical reaction - Process where two different elements are mixed together and creates a new chemical change to become a new substance.

Chromosomes- Thread-like strands that contain thousands of genes. Your genes determine the color of your eyes, height and other physical traits.

Chromatography - The chemical process in which components are separated.

Citric Acid - Acid that occurs naturally in fruits

Density - The density of an object is measured by the size of the molecules in the object, and how tightly packed the atoms are together with no holes. A cellphone has a high density because it is made from molecules that have a lot of mass. The atoms are tightly packed together with no holes.

DNA- Is short for deoxyribonucleic acid that is a molecule found in our cells which hold the secret to our genetic makeup.

Engineering - The process of solving problems and creating buildings, structures, and other creations by using math and science.

Fat - Fats are nutrients in food that the body uses to create energy. If too much fat cannot be used as fuel for the body, then they're stored in the body's fat cells.

Fluorescence - The light given off by a substance after it absorbs light from another source.

Freezing Point - The temperature needed to change a liquid to solid.

GAS - Substances like air that don't have a shape and move around freely, so you cannot hold it in your hand. Gas can be found everywhere from balloons to even soda.

Hygroscopic - Easy to obtain and hold moisture.

Molecules - Two or more atoms that are joined together.

Moon phases - Even 29.5 days the Moon completes a full cycle around the Earth. Each phase is a different position of the Moon's orbit. As the Moon orbits the Earth, the shadows from the Sun cast onto the Moon change our full visibility with each phase.

Non-Newtonian - Are materials that are slimy and don't have a shape just like liquids.

Osmosis - When water either enters or exits a substance.

Oxidation - Process where oxygen is combined with another substance to create a different compound.

PH balance - The pH of a substance measures how acidic or basic the element is on a scale of 0-14. The closer the number is to 0, the more acidic it is. The closer the number is to 14 the more basic it is. 7 is neutral.

Polymer - Long chains of molecules.

Quinine- A bitter liquid found in tonic water and other drinks. It is also used to treat Malaria.

Reversible Change - When a substance can be changed back to its original state after a reaction.

Saturation - When a solution is full and can no longer dissolve more than another element.

States of Matter – The three main states of matter are solid (has a shape), liquid (takes shape of container) and gas (invisible and fills container).

Static Electricity - Static electricity is when an electrical charge builds up on top of a static (unmoving) object over time. These electrical charges can be positive, negative or neutral.

Ultraviolet light (UV) - Short light waves produced by the Sun and other powerful sources of light. These waves of light cannot be seen with the naked eye.

Ventricles - Bottom chambers of the heart that squeeze and pump blood out of the heart.

Volume - The amount of space an object takes up.

Bonus Science Fun!

EXPERIMENT – WORD SEARCH

Help the scientists get ready to conduct the experiment.

T	M	S	L	A	C	I	M	E	H	C	V	Q	D	U	K	N	B	Y	F
W	O	I	V	M	X	E	B	T	L	A	B	C	O	A	T	J	B	O	D
B	T	E	O	M	I	K	K	Y	W	D	M	B	N	J	G	O	E	I	Q
R	E	N	R	U	B	N	E	S	N	U	B	P	J	L	Q	S	A	M	F
V	L	A	E	P	X	Y	K	V	A	P	J	G	T	V	O	S	K	J	O
W	Q	J	F	M	N	H	L	U	L	L	F	P	J	C	R	N	E	Z	Y
S	O	F	V	W	I	B	B	V	E	H	F	E	Q	A	I	E	R	F	G
H	Z	O	Y	V	D	R	X	H	I	Z	P	S	T	A	M	N	X	F	C
O	E	T	T	R	E	X	E	J	M	N	W	R	E	T	B	J	J	I	M
H	V	E	D	V	V	L	E	P	Y	K	O	Z	Z	L	E	Z	B	Z	T
O	O	U	D	H	S	W	A	Y	X	M	M	Y	C	M	G	P	F	F	W
M	W	R	G	L	A	S	S	E	S	E	G	I	Q	F	G	G	I	F	T
V	E	Z	H	Q	Q	V	V	P	T	M	I	C	R	O	S	C	O	P	E
D	X	S	S	E	B	U	T	T	S	E	T	W	I	R	B	W	I	G	S
V	S	R	S	G	W	V	I	V	J	P	H	E	S	F	J	O	I	V	M

Test tubes	Goggles	Glasses
Pipette	Beaker	Microscope
Chemicals	Bunsen burner	Experiment
Mortar	Lab coat	Flask

01

LABORATORY – DOT TO DOT

Connect the dots from 1 – 50 to complete the flasks and paint.

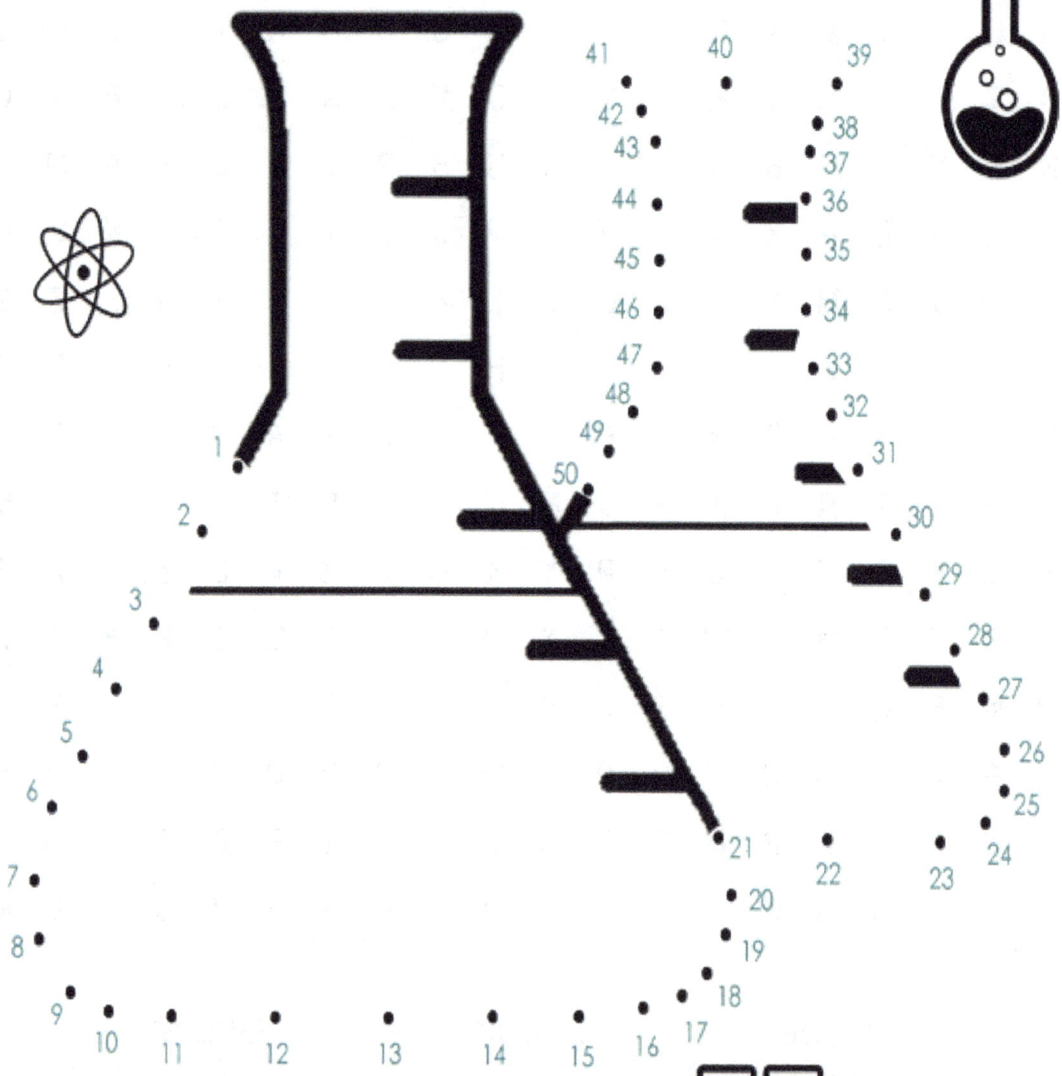

SPACE – CROSSWORD

Write the words and complete the crossword puzzle.

ACROSS

3. 

4. 

6. 

8. 

DOWN

1. 

2. 

5. 

7. 

SPACE – MAZE

Guide the rocket back to Earth through the maze.

SCIENCE – WORD SEARCH

Find and circle the words about science.

A	C	W	G	M	A	V	Q	B	E	R	J	K	Q	Z	V	W	K	J	C
X	P	E	R	I	X	F	I	E	O	N	J	T	W	V	Z	O	V	J	I
R	N	W	S	C	P	O	Y	H	C	A	N	A	T	O	M	Y	L	Q	T
C	F	T	U	R	L	M	O	T	Z	I	A	M	Q	P	T	J	T	Y	E
A	S	C	U	O	Z	S	N	H	I	L	G	J	E	U	F	P	A	J	N
K	R	Z	G	S	H	I	O	P	U	S	Z	O	J	Y	C	X	S	W	E
O	C	Y	X	C	D	N	J	D	S	V	N	L	L	N	Q	F	T	W	G
G	E	B	Y	O	M	A	F	E	G	C	H	E	M	I	S	T	R	Y	A
T	L	Q	R	P	Y	G	O	L	O	E	G	L	D	J	X	L	O	K	Z
G	L	J	B	E	P	R	Y	M	O	V	B	L	K	H	Z	J	N	W	L
L	K	X	D	S	R	O	R	P	H	Y	S	I	C	S	A	K	O	W	E
B	J	C	O	F	J	T	W	F	K	L	D	G	V	Z	X	E	M	S	U
O	X	W	Q	S	A	F	E	X	A	Z	L	Z	O	P	H	P	Y	C	Z
A	W	X	W	N	Q	F	Q	S	P	V	O	E	V	D	J	N	I	I	Q
X	J	P	L	W	V	S	P	O	Y	G	V	K	W	U	E	S	W	R	G

Astronomy
Chemistry
Geology
Cell
Logic
Anatomy
Microscope
Physics
Density
Organism
Genetic
Biology

SPACE – DOT TO DOT

Connect the dots from 1 – 60 to complete the rocket and paint.

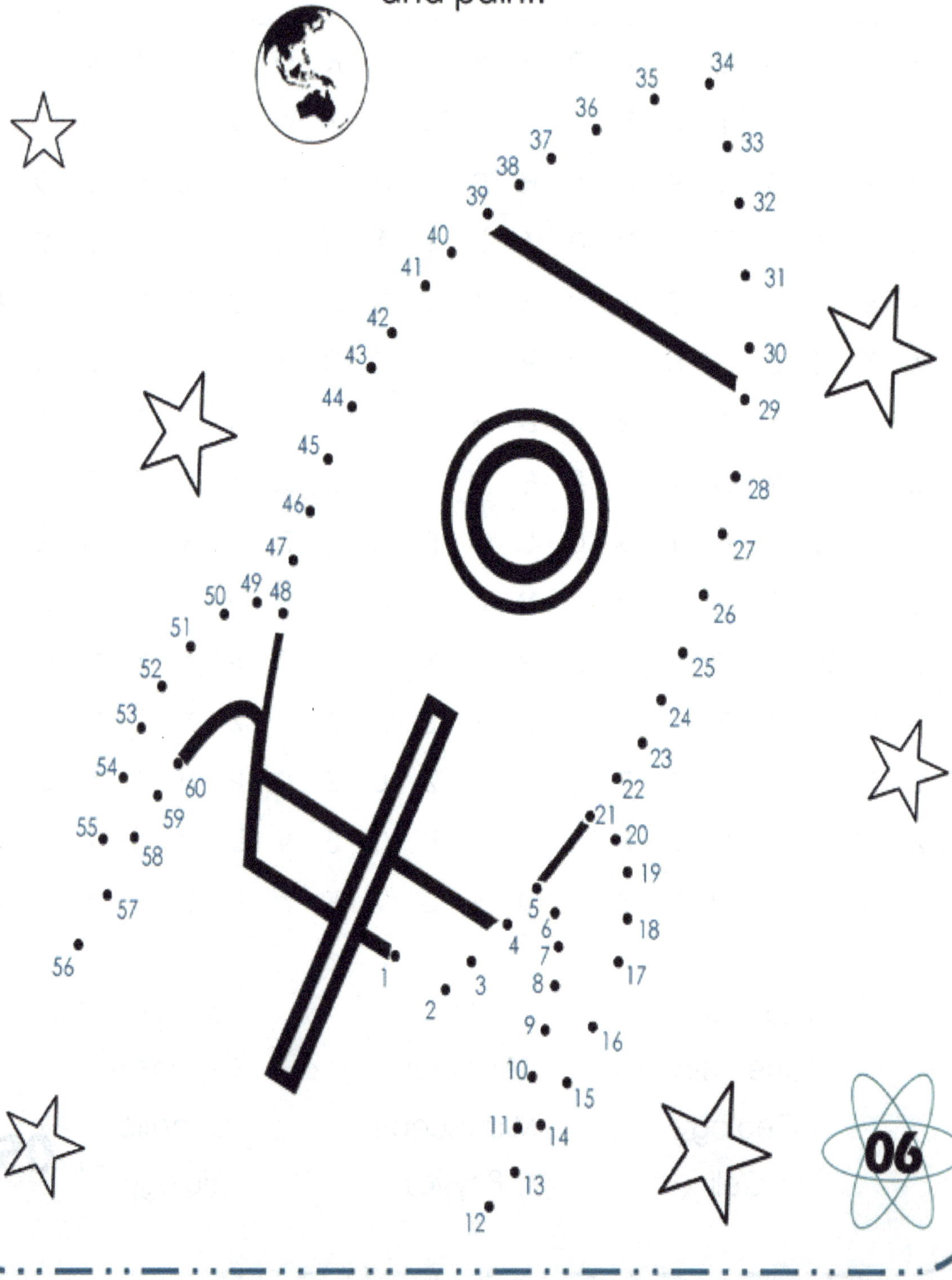

SCIENCE EQUIPMENT – CROSSWORD

Write the words and complete the crossword puzzle.

ACROSS

6. The piece of equipment used to measure temperature.
7. The equipment used to measure mass.
8. Used to separate solids from a liquid.
9. Used to hold solids or liquids, they come in different sizes.

DOWN

1. Used to transfer small volumes of liquid.
2. Used to hold equipment over a Bunsen burner.
3. Used to heat substances.
4. Sits on top of the tripod and spreads the heat from the flame.
5. Used to mix and heat chemicals.

LABORATORY – MAZE

Guide the scientist to the test tubes through the maze.

SPACE – WORD SEARCH

Find and circle the words about space.

C W
I X
D H Y L
L A T Q
G Y S F R A
N T T H D A
X C S U M C H T I R F P O E M G Q P N M
S Q H Z J M U B V O K X I O Q Z K L P P
A Q C Z A G P A N S J O R W F Q G L
I X J R K A R A X N C R M M H A
D J S E G G U A S H Q V K N
S D A T J T V T L W N E
H P L N E I N R A Q O T U F
Z A Q V N M P Z R M S R G S
S X C W Z O V O U S N C J B J Z
Y Y E T Y I F J T A O W I I
I A L F X B D A L S H J T Z
Y J B E Y U O J Z I
F J R H N W V D
O U W V

Astronaut	Earth	Jupiter
Galaxy	Mars	Stars
Orbit	Gravity	Moon
Space	Planets	Sun

LABORATORY – DOT TO DOT

Connect the dots from 1 – 50 to complete the mortar and paint.

SOLUTIONS

01

T	M	S	L	A	C	I	M	E	H	C	V	Q	D	U	K	N	B	Y	F
W	O	I	V	M	X	E	B	T	L	A	B	C	O	A	T	J	B	O	D
B	T	E	O	M	I	K	K	Y	W	D	M	B	N	J	G	O	E	I	Q
R	E	N	R	U	B	N	E	S	N	U	B	P	J	L	Q	S	A	M	F
V	L	A	E	P	X	Y	K	V	A	P	J	G	T	V	O	S	K	J	O
W	Q	J	F	M	N	H	L	U	L	L	F	P	J	C	R	N	E	Z	Y
S	O	F	V	W	I	B	B	V	E	H	F	E	Q	A	I	E	R	F	G
H	Z	O	Y	V	D	R	X	H	I	Z	P	S	T	A	M	N	X	F	C
O	E	T	T	R	E	X	E	J	M	N	W	R	E	T	B	J	J	I	M
H	V	E	D	V	V	L	E	P	Y	K	O	Z	Z	L	E	Z	B	Z	T
O	O	U	D	H	S	W	A	Y	X	M	M	Y	C	M	G	P	F	F	W
M	W	R	G	L	A	S	S	E	S	E	G	I	Q	F	G	G	I	F	T
V	E	Z	H	Q	Q	V	V	P	T	M	I	C	R	O	S	C	O	P	E
D	X	S	S	E	B	U	T	T	S	E	T	W	I	R	B	W	I	G	S
V	S	R	S	G	W	V	I	V	J	P	H	E	S	F	J	O	I	V	M

02

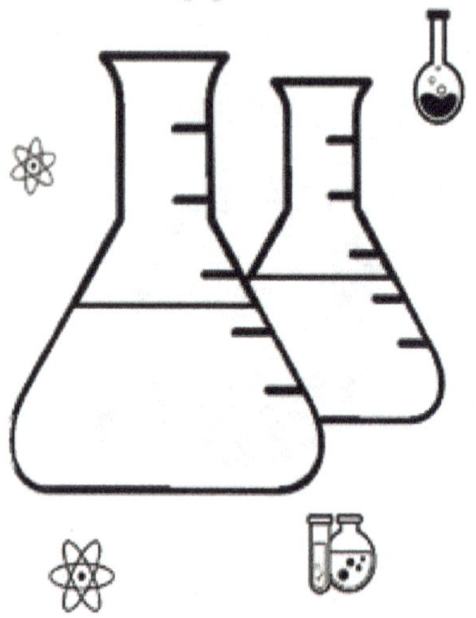

03

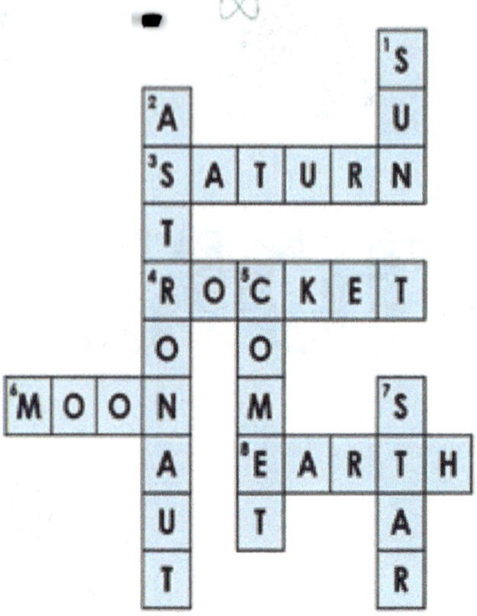

04

05

A	C	W	G	M	A	V	Q	B	E	R	J	K	Q	Z	V	W	K	J	C
X	P	E	R	I	X	F	I	E	O	N	J	T	W	V	Z	O	V	J	I
R	N	W	S	C	P	O	Y	H	C	A	N	A	T	O	M	Y	L	Q	T
C	F	T	U	R	L	M	O	T	Z	I	A	M	Q	P	T	J	T	Y	E
A	S	C	U	O	Z	S	N	H	I	L	G	J	E	U	F	P	A	J	N
K	R	Z	G	S	H	I	O	P	U	S	Z	O	J	Y	C	X	S	W	E
O	C	Y	X	C	D	N	J	D	S	V	N	L	L	N	Q	F	T	W	G
G	E	B	Y	O	M	A	F	E	G	C	H	E	M	I	S	T	R	Y	A
T	L	Q	R	P	Y	G	O	L	O	E	G	L	D	J	X	L	O	K	Z
G	L	J	B	E	P	R	Y	M	O	V	B	L	K	H	Z	J	N	W	L
L	K	X	D	S	R	O	R	P	H	Y	S	I	C	S	A	K	O	W	E
B	J	C	O	F	J	T	W	F	K	L	D	G	V	Z	X	E	M	S	U
O	X	W	Q	S	A	F	E	X	A	Z	L	Z	O	P	H	P	Y	C	Z
A	W	X	W	N	Q	F	Q	S	P	V	O	E	V	D	J	N	I	I	Q
X	J	P	L	W	V	S	P	O	Y	G	V	K	W	U	E	S	W	R	G

06

07

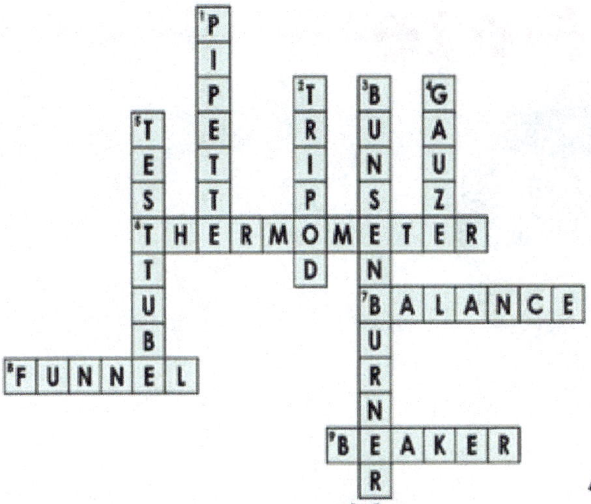
PIPETTE
TRIPOD
BUNSEN BURNER
GAUZE
TEST TUBE
THERMOMETER
BALANCE
FUNNEL
BEAKER

08

09

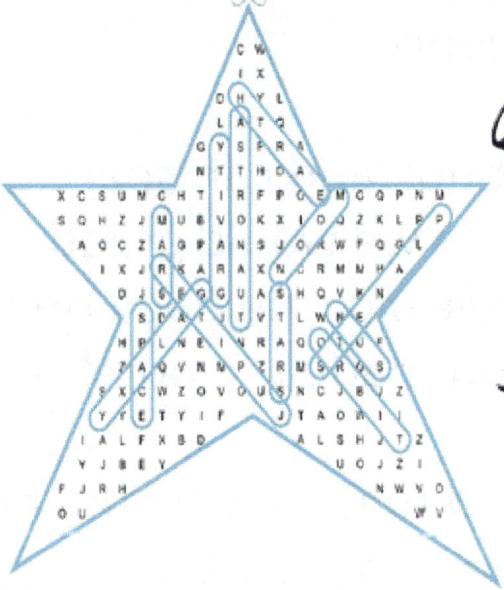

10

Final Note

We hope you have enjoyed the kitchen world of science.

If you enjoyed this book and can spare a moment of your time we would greatly appreciate your review of the book on Amazon or the site you made the purchase.

It would mean the world to us and would help others find the book as well.

About the Authors

The Wolf Cub Chlo team consists of Jenn aka Mama Wolf and Chloé aka Wolf Cub Chlo.

Jenn is a high school teacher, author, and mom. Literacy is her passion, but parenting is her joy. If she's not talking about marketing and business, you will find her humorously discussing teaching, restorative justice, parenting, and reading levels of children. She often discusses how much children are stuck to screens and either disdain reading or are struggling to catch up to their peers, so she and her daughter have made it their mission to find ways to bring the fun back to reading.

Chloé aka Wolf Cub Chlo is a young author who loves drawing, crafting, science experiments, reading, and telling silly jokes. Chlo wrote her first award-winning and best-selling book "Once a pun a Time- a guide to reading and telling jokes for kids" at the age of six. She was tired of seeing people sick, sad, and bored and wanted to find a way to use her brain to make more people smile.

Together, the Wolf Cub Chlo team encourages all to remember, that you are one smile away from helping re-shape the world.

www.wolfcubchlo.com

Other books by the authors

Book 1:
Edible Crafts:
A Family Baking Book for kids

Book 2:
Edible Crafts:
A Family CookBook
for kids

Once "a pun" a Time- A guide to reading and telling jokes for kids

Once "a pun" a Time- Artivity book for kids

Keep the party going with done-for-you bundle deals

As a thank you for purchasing the books, we are offering bonus bundle deals only for our readers.

These offers are only found in our books and are not being offered anywhere else. Check them out below

Nurture the budding comedian inside your child with the "Bundle of Laughs" Deal. This packed bundle includes a host of activities for children ages 5-9 years old. Activities range from a 50-page printable ART-ivity book and revised funny joke book for kids, to fun reading bingo games, punny game boards and more!

Visit https://jokes.wolfcubchlo.com/bundle

This packed bundle includes 4 Kitchen Science Activity Books for children ages 5-9 years old plus 4 surprise bonuses. Visit the link below to get done-for-you bundles of fun for a fraction of the cost!

Visit

https://science.wolfcubchlo.com/bundle

Made in the USA
Monee, IL
02 November 2026

67384740R00083